7 Step Action Plan

To Get LIVE Speaking Gigs with Chambers and Other Groups

By Maria Gudelis

Speak for Profits Now!

First of all, I am honored to give you the gift of knowledge that will allow you to achieve your financial goals. Thank you for purchasing this information product and I want to let you know if you have any comments or feedback, please let us know at Support@KeepItRealCoaching.com

We love hearing from our customers and treat you like gold!

Editorial Director: Angii Anderton

Cover Design: vMedio, Inc

Production and Composition: vMedio Publishing

© 2010 by Maria Gudelis
All rights reserved.
ISBN: 1449942865

For information about special discounts for bulk purchases, please contact vMedio Publishing, a division of vMedio, Inc at 519-331-7042 or visit www.MariaGudelisHelp.com

More great published books by Maria Gudelis:

- *21 Ways to Use Social Media*
- *The Twitter Business Advantage*

Manufactured in the United States of America

INTRODUCTION

I put together this <mark>quick 'take action' guide</mark> so that you get yourself in front of qualified business owners who will buy from you!

Combine this guide with the top secret tools here:

1. Live video presentation of me presenting at the Las Vegas Chamber:

   - **Video Streaming:**
     www.directmarketingsociety.com/coc
   - **Video Download:** (900 MB)
     www.tinyurl.com/maria-livepres-vid
   - **Audio MP3:**
     www.tinyurl.com/maria-livepresentation
   - **Ipod Video:**
     http://tinyurl.com/maria-ipodvid

2. The Last Chapter of the Live Video Presentation – 'Sales Hook' for selling social media marketing

3. **The** Blueprint Questionnaire you give them at end of the presentation, **go here:**
   www.tinyurl.com/maria-bluprintQ

4. **PowerPoint presentation** you can swipe it and customize for yourself! Go here: www.tinyurl.com/maria-chamberPPT

Note:

You don't have to limit yourself to Chambers of Commerce as there are so many other organizations, associations, meet up groups you can present to.

READY TO GO?

You bet!

Okay, before diving into the 7 steps…let me tell you the story of how I got in front of local business owners at the Chamber and also…other nifty groups that fits your target market.

First of all, it just didn't happen 'overnight'. You know how sometimes you have to go through a lot of 'no's' before getting that coveted 'yes'?

Know this: you very well could have to talk to over ten people to get your first workshop scheduled…

Would that be worth a $1,000 or $5,000 plus payday?

You bet!

If you are shy to call or scared…SO WHAT! What will really happen once you make that first call, email or oh my goodness…one to one 'in person' contact with the person in charge of selecting speakers…

The best outcome is getting your workshop scheduled…the worst…they say no or put you on a 'waiting' list in case one of their existing scheduled speakers gets sick/has to cancel and you pop up as a backup!

I beg you to be persistent at this and it will pay off in diamonds for you and it is like riding a bike, once you get the hang of it…it comes oh so natural!

PERSISTENCE PAYS OFF!

Did you know? Richard Bach wrote a story about a seagull and was turned down by 18 publishers…yes 18! And then…when  McMillan published his story…

Jonathan Livingston Seagull sold more than 7 million copies within 5 years!

Successful entrepreneurs realize rejection is just part of the process in achieving their goals, their visions…I want you to be a successful entrepreneur and never give up, never quit – feed your energy with the no's…persevere to your greatness!

MY STORY

So getting back to my story,

My 1st action: I first contacted the Las Vegas Chamber of Commerce by phone and asked:

> *"Hi, my name is Maria Gudelis, I was wondering if you could help me?*

The person answering the phone will say yes or how can I help you.

> *"Who may I speak to in regard to being a speaker for one of your business workshops you offer to your members?"*

They then gave me the name of the individual in charge and…proceeded to tell me that, *"all their speakers for workshops are already booked for the next year."*

Yes, a whole year out! This is serious business! The Las Vegas Chamber is a huge business, over 7,000 members and one of the largest chambers of commerce organizations in the world!

So if one of 'us little guys or gals' can break into this Chamber – you can too in your city!

So I then proceeded to call this person in charge of bestowing the mighty honor of being a guest speaker for a Chamber workshop.

My 2nd Action: Ring, ring…got a voicemail, left a message with this exact script:

> "Hi (name of person in charge of selecting speakers), my name is Maria Gudelis and I am an expert in social media consulting, specifically online video marketing and twitter and would love to be a guest speaker for the chamber at no cost to you. Online video and twitter marketing tactics can help your members increase their profits, please call me back at your convenience at 702-xxx-xxxx…

No call back…

My 3rd Action: Ring, ring…got a voicemail, left a message very similar to above script.

3 weeks later: Okay, I figured I have absolutely nothing to lose now and decided to call the manager of 'member services'. I explain that I am a social media expert and have been trying to get in touch with (name

of person), and have been disappointed in not hearing back from her.

One day after that call…

My phone rings. It's the person from Chamber in charge of selecting speakers:

> *Hi Maria, this is (name of person), and I really apologize for you not getting my return calls, how can I help you?*

I say:

> *Hi there, thanks for getting back to me…well I would rather like to turn that question around (first name of person) and ask you, how can I help the Las Vegas Chamber of Commerce and your members as I can teach a variety of great topics as it relates to marketing on the Internet…and…*
>
> *Since you are talking to your members who are business owners every day…what topics do you think they would get the most value from?*

• • •

Side Note from Maria:

THIS IS HUGE – did you notice how I turned the conversation around?

Write this down now as one of your HOT Action Tips: Ask how YOU can help them…turn it around from get to give situation.

Also, when you are asking questions, you are then CONTROLLING the conversation.

• • •

Person from Chamber in charge of selecting speakers:

> *Well Maria, now that you mention it…I really thought social media and twitter was already a covered topic as we have already had speakers on this.*
>
> *However, our members are requesting more workshops on twitter specifically, I also love the topic of video marketing as no one has taught about online video marketing yet. If you can 'dumb' Twitter down for us…as in recognize our members are mostly, well older, and some of the past speakers went way over our heads, I'd like to schedule you in.*

I say:

> *(name of person), I'd be glad to 'dumb Twitter' down for the presentation and if you will permit me, I'll make the video marketing part so easy, I'll even bring my flip video camera with me to show everyone that it is such a simple process to push a big red button versus all the video camera's we're used to seeing, even my grandma can use a Flip video camera!*

Person from Chamber in charge of selecting speakers:

> *Great Maria, well I have these two dates available, which one works better for you and can you send me a brief paragraph describing your presentation so I can modify it as per chamber guidelines and get it out into our newsletter.*

**** And that's a wrap! ****

Moral of the Story

Perseverance Pays Off and don't be scared to complain or 'go higher' if you aren't getting the service you desire...you know the saying 'the squeaky wheel gets the grease"!

You see, it really is simple to do…that is how I got my first Chamber presentation workshop and out of that exact workshop, I got two joint ventures (one of them is with a big accounting firm).

Feel free to use the same scripts above if you want to and simply start contacting your local chambers, associations and meet up groups!

Now I'd like to go over the seven step action plan for you.

So that you can be 'a wanted speaker' anywhere, anytime…

…Imagine yourself travelling at will, anywhere in the world…

…all you need is a laptop! ☺

Step 1: Position

Before you start contacting anyone, you need to figure out how you will be positioning yourself. Determine your 'title' and what you will position to others what you are expert at.

People love the word expert – use it – own it – practice it yourself in front of a mirror and with friends and family if you aren't confident YET in saying it…so say it out loud to the world:

I am a _____ expert!

(put in the blank what you decided – e.g social media, twitter, internet marketing, direct response, online video marketing, etc.)

In my case, I leveraged the fact that 'social media' is the big hot button right now to use in local business circles and said I was a 'social media expert'. I tied that into my business cards and title:

Maria Gudelis,

President of vMedio, Inc

Keep It Real Media

Whatever you decide, just make sure you are consistent and confident in your communications.

Step 2: Target

Here is the easy part as now with the internet, it is oh so easy to find so many organizations who would love to have you present.

So get yourself a cup of java or tea and take just 30 minutes to get 20 qualified leads to call.

Now Get This: All you need to do is go to Google and find:

1. Your local chamber of commerce and their contact info online

2. Your local meetup (www.meetup.com) organizations that will be glad to have you present

3. Your local associations that are begging for people like you to present.

Your mission: Write down from the 3 sources above 20 qualified leads you will target.

Need help? I've included a few screen shots to show you just how easy this is to target:

Screenshot One – This illustrates the chamber directory where you can click on state or go international to find the website and phone number of each chamber.

www.uschamber.com/chambers/directory

Now when I click on **Nevada**, I get:

Chamber Name: Henderson Chamber of Commerce
Address: 590 South Boulder Hwy
City: Henderson
State: NV
Zip Code: 89015
Phone Number: (702) 565-8951
Fax Number: (702) 565-3115
Website: www.hendersonchamber.com

Chamber Name: Humboldt County Chamber of Commerce
Address: 30 W Winnemucca Blvd
City: Winnemucca
State: NV
Zip Code: 89445
Phone Number: (775) 623-2225
Fax Number: (775) 623-6478
Website: www.humboldtcountychamber.com/

Chamber Name: Las Vegas Chamber of Commerce
Address: 3720 Howard Hughes Pkwy
City: Las Vegas
State: NV
Zip Code: 89109
Phone Number: (702) 641-2614
Fax Number: (702) 735-2011
Website: www.lvchamber.com

And this is a beautiful thing as I can now call and even check out their website.

Advance tip for offline consultants:

Some of you have taken my coaching and know one of my secrets…I have my VA contact business owners from those online directories with my email profit pulling scripts…now that is a million dollar tip!

Now let's go to the great target resource of local meet-up groups!

Go to **www.MeetUp.com** and

➔ Click on the FIND a Meetup Group.

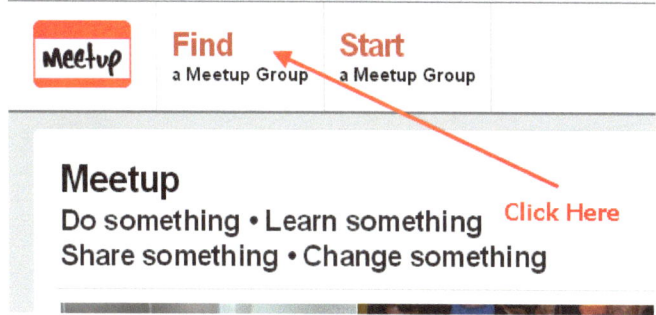

On the next screen you will see:

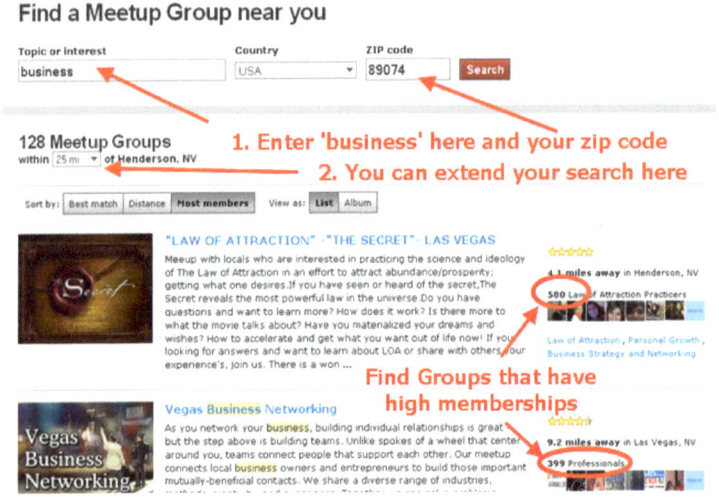

➜ Now click on the group(s) you are interested in contacting and you will see the next screen **with their contact info!** Cool huh??

Now let me tell you a bit about the third way to get target leads for speaking gigs…associations and clubs.

All you have to do is google 'your city' associations and 'your city' clubs and voila – you have a huge list to contact.

Step 3: Contact

Whew, okay, breathe deep now…here is the toughest part, you have to contact them! Yes you heard right and shoot if you can't call them on the phone and send them an email…then buck up and get some confidence, you have greatness within you and other business owners need your internet marketing services.

You are doing a disservice to yourself and the world by not sharing your talent, your knowledge!

I harp on that a bit as I have so many students who were terrified to make that 'first contact' but once they did…it was like smooth sailing after that and their bank account got bigger as a result.

You want that too don't you?

So get on the phone, hop into email and start contacting! Use the similar script I have given to you in the "My Story" section or use your own style.

The biggest point I need to tell you is this:

Take action!
You'll shock yourself
with the results

Step 4: Send Info

Congratulations! You've done the hardest part, step 3 and now step 4 is super easy.

All you have to do now is this: Send them info on your topics and don't forget to highlight the BENEFITS to them…What's In It for Them. Oh, and phone them to confirm they received your email with that content!

At least in my experience this is a necessary action after contacting and getting someone who:

1. Is interested in scheduling you to speak to their group.

2. Needs you to send in a brief description of the workshop so they can blast it out to their members and pack a room for you!

So to help you on this, I've included the exact same description the Las Vegas Chamber used for my event so you can use as your swipe file if you wish.

Feel free to use your own ad copy, description as well!

● ● ● ● ● ● ● ● ● ● ●

Business Survival:
Guerilla Marketing Tactics, Video & Social Media

How to Use Simple Video Social Media Tactics and Boost Your Revenue

Member Speaker Maria Gudelis, President of vMedio, Inc. Keep It Real Media, will instruct you how to use simple video social media tactics and boost your revenue.

Discover low cost, fat profit tactics using a simple $120 Flip Video Camera and "free social media" that are so downright easy anyone can do this! This is the fast way for small businesses to ATTRACT CUSTOMERS.

The 1.5 hour presentation will focus on the flip video camera, how easy it is to use and leverage the power of free social media sites to attract new customers. That includes a very basic dumbed-down discussion about Twitter, and then really zeroing in on video – it isn't as hard or expensive to use as most business owners think!

There is no charge for Las Vegas Chamber of Commerce members, $20.00 for non-members to attend.

Seating is limited so reserve your seat now.

To register please click on the *"Register Now"* button or call the Chamber at (702) 641-5822 and ask for Member Services.

Deadline to register for this event is May 15, 2009.

Hosted by:

*   *   *   *   *   *   *   *   *   *   *

And here is a screen print of what they put on their website and blasted out to their members:

Event Details

Event Calendar | Community Calendar

Business Survival: Guerilla Marketing Tactics, Video & Social Media

. Please Note --- This is a Past Event!! .

How to Use Simple Video Social Media Tactics and Boost Your Revenue

Member Speaker Maria Gudelis, VP of Social Media, Wildhorse Performance Marketing will instruct you how to use simple video social media tactics and boost your revenue.

Discover low cost, fat profit tactics using a simple $120 Flip Video Camera and "free social media" that are so downright easy anyone can do this! This is the fast way for small businesses to ATTRACT CUSTOMERS.

The 1.5 hour presentation will focus on the flip video camera, how easy it is to use and leverage the power of free social media sites to attract new customers. That includes a very basic dumbed-down discussion about Twitter, and then really zero'ing in on video – it isn't as hard or expensive to use as most business owners think!

There is no charge for Las Vegas Chamber of Commerce members, $20.00 for non-members to attend.

Seating is limited so reserve your seat now.

To register please click on the **"Register Now"** button or call the Chamber at (702) 641-5822 and ask for Member Services.

Deadline to register for this event is May 15, 2009.

Hosted by:

Las Vegas Chamber of Commerce

Need more information
If you need more information about this event, please complete fields below:

Your Email Address:

Your Name:

Question / Comment:

Send ☒

For general inquiries e us at: info@lvchamber

Step 5: Close

Now some of you might think this is the hardest part because of the word 'close'!

However this is the easiest because you've mastered a 'relationship' with that person you contacted and now is it just a matter of getting you scheduled in on a date that works for you and for them.

Simply follow up to confirm a date and seriously, 100% of the time, they are calling you to confirm the date – that is how HOT this market is right now for your services!

Seriously, you can contact my Performance Marketing Manager, Trish, and she can tell you how often they have called, called and called me to book the day and time!

Step 6: Prepare PPT

Those who plan, plan to succeed. Those that don't plan, plan to fail. So don't just grab my lovely power point slides I've prepared for you and present without reviewing it!

So go through your presentation, customize it for your company, your needs and decide if you will go 'live' using a live internet hook up and what websites or 'exercises' you might do with them.

Sometimes if time is allowed, I do that – go to the internet to display examples or even ask someone in the audience 'Who wants a FREE website consultation right now?'

And I go to their website and tell them what is 'broken' and how to fix it. 100 percent of the time, they don't have such basic things such as a:

1. Call to action

2. Autoresponder (customer acquisition and retention system is what I call it to the business owners as most don't know the term 'autoresponder')

3. Video on their site

4. Free report or something to entice a visitor to give up their email ID

5. SEO – now keywords…(*funny story for you: one time I did this and their keywords when I did a View-Source to show everyone in the room the source code to find the keywords….well, this business owners keywords was one keyword: xsitepro! No kidding – this was a perfect 'segway' into explaining why most graphic designers you hire know nothing about SEO or direct marketing online!*)

6. Good graphics and marketing colors (yes – one time one of my prospects had actually 'PINK' text on here site instead of black text – ouch on the eyes!)

7. Benefits identified about their product or service. This is a biggie, usually it is all ego – about them, their company. Who cares, the potential customer coming to their site wants to know what's in it for them. Period. Gone in 5 seconds otherwise!

If you are seriously interested in learning more on the 'Website Repair Tactic", I have a product that details step by step how to get paying clients fast.

All you have to do is go here to check it out: **www.WebsiteRepairTactics.com**

Step 7: Present &Hook

Step seven is the fun part. Now you present! A few things to note:

Side Note from Maria:

You can get yourself a free copy of a media release form by entering 'media release form' into Google.

1. Never ever give printouts of your presentation to the audience. Why? Well then they may be given 'too much' information so they don't call you!

2. Relax, have fun with this!

3. Be professional in your attire, dress business top notch or casual – depends on where and what group you are presenting to. Remember, it is always better to be overdressed than under dressed. For the ladies, go easy on the jewelry, keep it simple.

4. If you are allowed to video tape and also record the audience…then bring 'media release forms' for participants to sign. This is a great segway into:

5. Get video testimonials right away while everyone is so excited about what you taught!

6. Make sure you give out your blueprint questionnaire as detailed in the video segment chapter 7 of my video presentation to you.

Some of the information I shared in chapter 7 is worth repeating here:

*** TRANSCRIPTION OF SEGMENT 7 ****

I am going to tell you how to end your workshop to be most effective to get clients. The first thing you need to do is this:

You have a goal that you want to qualify, that's why I have it written right out, you want to qualify everyone you just presented to, right?

How do you do that?

One of the best ways to do that is to give them a questionnaire, and I like to give them a questionnaire for them to fill out at the end of the presentation and give them a hook, like a rule, a prize. You know, everyone wants something back, right? They want something if free. So what's the hook if they fill out that questionnaire for you?

The hook is a free $247 gift.

And guess what that gift is, a one on one consult.

Now, if you presented to say 10 or 20 business owners, how do you then close them?

How do you then get them to give you a check, for you to do marketing services?

Well, the easiest way to do that is then to have, you've already **gained know- like- trust**, you know the KLT factor. And what you want to do is then now that they **know you and feel they can trust you** and they know you're the expert because you have presented the PowerPoint presentation that I've even given to you as well, **that puts you up there as an expert**.

All you need to do now is basically call them, or ask them to come to your office for 15 minutes. That's about it. And you're going to give them some more free advice how you're going to help them one on one with their business, and here's the thing is that, on the Questionnaire you gave them at the end of the workshop to claim their free consult... guess what they have to fill out:

> *"How much do you spend on marketing or advertising per month?"*

> *"Are you satisfied with the results that you're getting?"*

And then I ask:

> *"How much are you spending on offline marketing?"*

You want to know how much they spending on the yellow pages.

Because, imagine this: If I am presenting in front of 20 business owners, it's hard to ask everyone individually in the room how much they're spending, right?

Maybe you could get an answer from one or two of them, but you don't want to ruin the flow of your presentation qualifying them that way right then and there.

So, qualify with a questionnaire and provide them a gift, that hook of free time with you, a 15 minute consult worth $247, and let's face it, the 15 minutes could go half an hour.

And, if you get on the questionnaire that they are not spending any money on marketing, well you sort of qualified your lead, right?

Out of 20 questionnaires that you get, if you see one that spending 10,000 a month on marketing, and one spending $100 a month on marketing, you kind of know who you want to target, right?

So, that's the benefit of closing your workshop in that manner, and make sure you also have your business cards handy. You leave them at the back of the room and say, also look guys if you want to have knowledge of how to contact me, I hope I've given you enough information now that you could really go out and use the free sites of social media, and really pump up your profits online now, so you could spend more time with family.

What I'd like you to do if you're stuck anywhere or have any questions, grab my business card, I'll be glad to talk to you, just give me a call and then guess what everyone:

On your business card, have a call to action on the back of the card, have a call to action, reiterating the "one on one consult".

So, on it say – "Get your free risk free 'One on One Consult'" and you could even say $247 value.

So, that's really it, and I hope you enjoyed watching the entire presentation that was presented at the chamber of commerce in the Las Vegas, Nevada, and that you will go out yourself and do workshops whether it'd be live in front of business owners, whether it'd be virtually setting up a go to webinar meeting and increase the revenues for your consulting company this year now.

Note:

The 'virtual client getting' is HUGE, one of my students took the info I gave him, got 27 business owners on the virtual webinar and got 3 appointments booked. Another student got 10 business owners on a virtual webinar and got a $1,000 check from it!

So what are you waiting for?

The speaker microphone is waiting for you now…all you have to do right now is take action on the seven steps you've been given in this guide and grab that microphone to profits!

IF YOU WANT TO KNOW MORE INFORMATION ABOUT PERSONAL COACHING OR MENTORING PROGRAMS FROM MARIA GUDELIS,

PLEASE CALL OUR OFFICES NOW AT 519-331-7042

OR CHECK OUT WWW.MARIAGUDELIS.COM NOW!

Free CD Offer

Let me send you my CD:

"Simple Social Media Trends to Boost the Earning Power of Your Business"

All you need to do is get online and go to:

www.vMedio.com/freecd

---

www.ingramcontent.com/pod-product-compliance
Lightning Source LLC
Chambersburg PA
CBHW040930180526
45159CB00002BA/678